Get to Know Big Cats

TIGERS

By Bray Jacobson

Please visit our website, www.garethstevens.com. For a free color catalog of all our high-quality books, call toll free 1-800-542-2595 or fax 1-877-542-2596.

Library of Congress Cataloging-in-Publication Data
Names: Jacobson, Bray, author.
Title: Tigers / Bray Jacobson.
Description: Buffalo, New York : Gareth Stevens Publishing, [2024] | Series: Get to know big cats | Includes index. | Audience: Grades K-1
Identifiers: LCCN 2022045082 (print) | LCCN 2022045083 (ebook) | ISBN 9781538286142 (library binding) | ISBN 9781538286135 (paperback) | ISBN 9781538286159 (ebook)
Subjects: LCSH: Tiger–Juvenile literature.
Classification: LCC QL737.C23 J3375 2024 (print) | LCC QL737.C23 (ebook) | DDC 599.756–dc23/eng/20220920
LC record available at https://lccn.loc.gov/2022045082
LC ebook record available at https://lccn.loc.gov/2022045083

First Edition

Published in 2024 by
Gareth Stevens Publishing
2544 Clinton Street
Buffalo, NY 14224

Editor: Kristen Nelson
Designer: Leslie Taylor

Photo credits: Cover, p. 1 dangdumrong/Shutterstock.com; pp. 5, 7, 11, 24 (whiskers) Volodymyr Burdiak/Shutterstock.com; p. 9 Julian W/Shutterstock.com; p. 13 TigerStocks/Shutterstock.com; p. 15 GUDKOV ANDREY/Shutterstock.com; p. 17, 24 (claw) Suntisook.D/Shutterstock.com; p. 19 Danny Ye/Shutterstock.com; pp. 21, 24 (cub) Anan Kaewkhammul/Shutterstock.com; p. 23 Anuradha Marwah/Shutterstock.com.

Printed in the United States of America

CPSIA compliance information: Batch #CSGS24: For further information contact Gareth Stevens at 1-800-542-2595.

Contents

Tigers are big cats!

They have fur.
It is striped.

Fur can be yellow.
It can be orange.
It can be white!

Tigers have whiskers. These help tigers move in the dark.

Tigers hunt alone.
They hunt at night.

They eat deer
and pigs.
They eat cows
and horses.

Tigers have long claws.
They are used to hunt.

Tigers have webbed paws. They help tigers swim well.

Babies are cubs.
They cannot see
at birth.

They stay with
their mother.
They learn to hunt!

Words to Know

claw

cub

whiskers

Index